LET'S FIND OUT! BIOMES

WHAT IS THE TUNDRA?

JULIA J. QUINLAN

Britannica Educational Publishing

IN ASSOCIATION WITH

ROSEN EDUCATIONAL SERVICES

Published in 2019 by Britannica Educational Publishing (a trademark of Encyclopædia Britannica, Inc.) in association with The Rosen Publishing Group, Inc.
29 East 21st Street, New York, NY 10010

Distributed exclusively by Rosen Publishing.
To see additional Britannica Educational Publishing titles, go to rosenpublishing.com.

First Edition

Britannica Educational Publishing
J.E. Luebering: Executive Director, Core Editorial
Mary Rose McCudden: Editor, Britannica Student Encyclopedia

Rosen Publishing
Amelie von Zumbusch: Editor
Tahara Anderson: Series Designer/Book Layout
Cindy Reiman: Photography Manager
Nicole DiMella: Photo Researcher

Library of Congress Cataloging-in-Publication Data

Names: Quinlan, Julia J., author.
Title: What is the tundra? / Julia J. Quinlan.
Description: New York : Britannica Educational Publishing, in Association with Rosen Educational Services, 2019 | Series: Let's find out! Biomes | Audience: Grades 1–5. | Includes bibliographical references and index.
Identifiers: LCCN 2018016340| ISBN 9781508106999 (library bound) | ISBN 9781508107132 (pbk.) | ISBN 9781508107347 (6 pack)
Subjects: LCSH: Tundra ecology—Juvenile literature. | Tundras—Juvenile literature.
Classification: LCC QH541.5.T8 Q85 2018 | DDC 577.5/86—dc23
LC record available at https://lccn.loc.gov/2018016340

Manufactured in the United States of America

Photo credits: Cover and interior pages background jjsupasit srisawthsak/Shutterstock.com; p. 4 Nichole Andler /U.S. National Park Service; p. 5 © Encyclopædia Britannica, Inc.; p. 6 U.S. Fish and Wildlife Service; p. 7 Bob Wick/U.S. Bureau of Land Management; pp. 8, 14 Pi-Lens/Shutterstock.com; p. 9 Bertrand louis/Shutterstock.com; p. 10 robert cicchetti/Shutterstock.com; p. 11 Denny35463/E+/Getty Images; p. 12 © amarinchenko106/Fotolia; p. 13 © Patrick Morrow/First Light; p. 15 Incredible Arctic/Shutterstock.com; p. 16 © AdstockRF; p. 17 Ana Flasker /Shutterstock.com; p. 18 © pilipenkod/Fotolia; p. 19 AndreAnita/Shutterstock.com; p. 20 Yobab/Shutterstock.com; p. 21 Gary Gray/iStock/Thinkstock; p. 22 Michael Sewell/Photolibrary/Getty Images; p. 23 Nadia Isakova /AWL Images/Getty Images; p. 24 Denis Kozhevnikov/TASS/Getty Images; p. 25 Roofsoldier/Shutterstock.com; p. 26 Milan Sommer/Shutterstock.com; p. 27 Sebastian Kennerknecht/Minden Pictures/Getty Images; p. 28 Leonard Lee Rue III/Science Source; p. 29 Jasper Doest/Minden Pictures/Getty Images.

Contents

Welcome to the Tundra

All the land on Earth can be divided into biomes. Each biome is a large region that has a certain climate and certain types of living things. The tundra is one kind of biome. A tropical rainforest is another example of a biome. The tundra is very different from a tropical rainforest, though!

Musk oxen graze on the tundra in Bering Land Bridge National Preserve in western Alaska.

The tundra is a large, **barren** region with no trees. Large areas of tundra lie between the permanent ice of the far north and the northern forests of North America, Europe, and

Asia. They cover about 20 percent of Earth's surface. They are some of the coldest places on Earth.

There are two types of tundra. They are the Arctic tundra and the alpine tundra. Arctic tundra lies in northern Europe, Russia, Alaska, Canada, and Greenland. Alpine tundra is found high on mountaintops, where trees cannot grow.

VOCABULARY

In a **barren** region, there are not many living things.

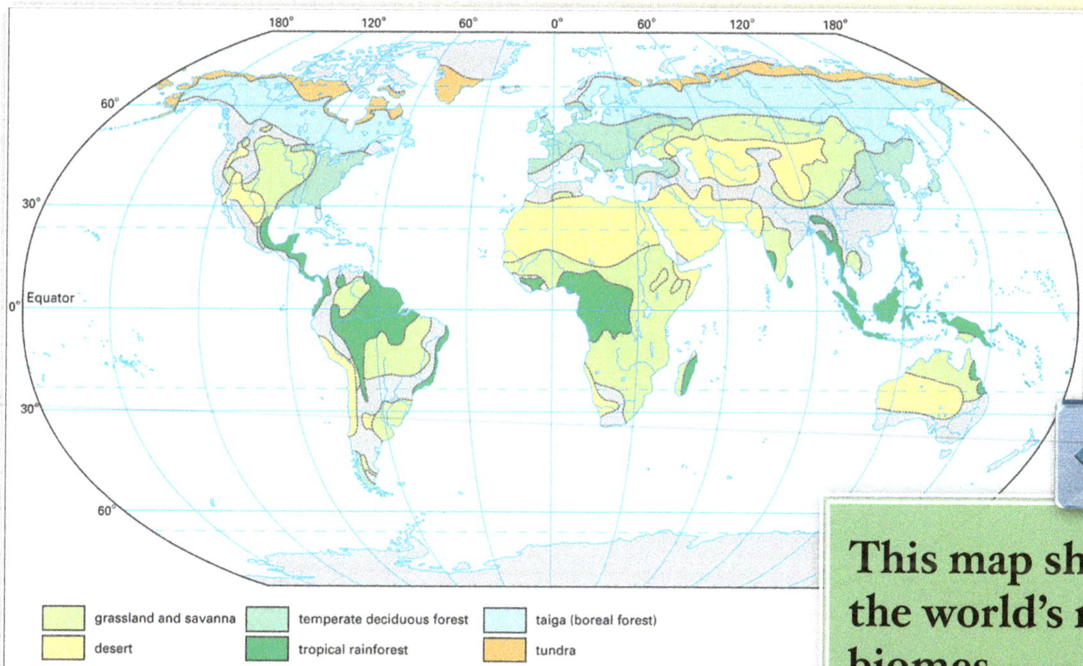

This map shows the world's major biomes.

The Coldest Biome

The tundra is a cold, dry biome. It may be flat, hilly, or mountainous. Little plant life grows there. Tundra in coastal areas tends to be foggy. Snow covers the tundra for more than six months of the year.

In the Arctic tundra, the winter temperature may be as low as -25 degrees Fahrenheit (-32 degrees Celsius). The summer temperature may rise to only 40°F (4°C). The ground experiences a constant cycle of freezing

Alpine tundra is often mountainous, like this area in Alaska.

Think About It

The sun's heat causes water to evaporate, or to change from a liquid into a gas. Why do you think there is little evaporation in the Arctic tundra?

in winter and thawing in summer. Because of the cold temperatures, the Arctic tundra has a permanent layer of frozen soil, called permafrost, under the surface. Some permafrost reaches as deep as 1,500 feet (457 meters). Because of the permafrost, little drainage takes place in the Arctic tundra. Since there is not much evaporation, what water there is tends to collect in bogs and ponds.

The Alaska Range looms over the bogs of Alaska's Delta Wild and Scenic River Watershed.

Arctic tundra covers much of the land between the ice-covered areas around the North Pole and the taiga. The taiga is a forest biome. The main feature of the taiga is its conifer forests. Conifers are trees that form seeds inside cones. Most conifers are evergreens, or trees that keep their needles year-round.

Alpine tundra lies farther south than the Arctic tundra. Alpine tundra is found in high

This section of taiga is in Yukon. Both taiga and tundra biomes can be found in this Canadian territory.

mountains above the tree line. (The tree line is the highest place where trees can grow.) It has short, cool summers and less extreme winters than the Arctic tundra. Alpine tundra does not have a layer of permafrost.

COMPARE AND CONTRAST

What makes the alpine tundra different than the Arctic tundra? What are the similarities? Why are they both classified as tundra?

This alpine tundra is located high in the Alps of Italy, above the tree line.

Tough Weather

The Arctic tundra receives less than 15 inches (38 centimeters) of rain every year. However, it can snow between 25 inches (64 cm) and 75 inches (191 cm) in the Arctic tundra. In the alpine tundra, precipitation can be up to 25 inches (64 cm) in the Rocky Mountains of Colorado but

An alpine tundra in the Rocky Mountains has a more moderate climate than the Arctic tundra.

may be less than 3 inches (7.6 cm) in the Himalayas, a mountain range in India, Nepal, Bhutan, and China. Blinding snowstorms make it difficult to see during the winter months, and summer rains can be heavy.

Animals that live in the tundra, such as these elk, must be able to survive in harsh weather.

Blizzards in Arctic and alpine tundra may make it hard to see farther than about 30 feet (9 meters). The winds blow snow crystals through tiny openings in clothing and buildings. Winds in the alpine tundra are often quite strong as well. They can reach 75 to 125 miles (120–200 km) per hour.

COMPARE AND CONTRAST

How are blizzards in the tundra like those in other biomes? How do you think they would be different?

What Grows in the Tundra?

Only low-growing plants, such as mosses and shrubs, can survive in the tundra. Plantlike living things called lichens also grow there. Most of the plants are a greenish-brown color. Plants do not flower for more than a few days or weeks, but these blooms can be large and

Orange lichen grows on a rock with various plants.

Think About It

Tundra plants tend to be small and low to the ground. What do plants need to grow? Why do you think the plants are so small in the tundra?

colorful, especially in alpine tundra.

Tundra near coastal areas can be foggy and have mossy or grassy landscapes. There, fungi and algae can grow. In the southern parts of the Arctic tundra, there are many bogs, lakes, and rivers. Grasses and

Cotton grass and mosses grow in the Arctic tundra in Yukon, Canada.

mosses grow here, as well as low willow trees and even some plants in the sunflower family.

The types of plants that grow in alpine tundra depend a lot on the elevation and soil. On gentle slopes, where soil has developed, meadows can form and expand. In other areas, like on higher mountains with snow and ice, lichens and mosses manage to grow on rocks.

These cloudberries are growing in the alpine tundra of Labrador, Canada.

Most plant growth in the tundra is limited to the summer months, when the temperature is above freezing. This can be as little as two months of the year. Plants in the tundra have evolved to be able to withstand this tough climate. They flower quickly and some can survive the tough winters.

VOCABULARY

Living things have **evolved** by changing slowly over many years.

This tundra flower is called purple saxifrage, and it is edible!

Amazing Animals

Just like its plants, the animals of the tundra are adapted to their harsh environments. There are fewer animals in the tundra than in other biomes.

Many animals in the tundra are migratory. This means they move from place to place during the year. They live in the tundra during the warmer seasons but leave when it becomes too cold. Caribou migrate between the tundra and taiga. They give birth to their calves in the tundra when plants are

An Arctic fox's thick fur helps it survive during the cold winters on the Arctic tundra.

Compare and Contrast

Some animals in the tundra come out of hibernation for only a few months of the year. How is that similar to plants that only bloom for a short time? How is it different?

in full bloom. They retreat to the taiga for the winters.

Other animals live in the tundra year-round. Many of these animals have thick layers of fat and thick, warm coats. Polar bears, snowy owls, Arctic wolves, and musk oxen stay in the tundra all year. Arctic ground squirrels also stay year-round. They live in large colonies and hibernate seven to nine months of the year.

Some caribou, also known as reindeer, live in the Arctic tundra.

ARCTIC ANIMALS

Animals living in the Arctic tundra must be able to live through freezing cold winters and also reproduce during the short summer months.

Some Arctic animals have coats that change color with the seasons. Their coats turn white for the winter, making them blend in easily with the snow. In the summer, they turn brown. Animals that have color-changing coats include the Arctic fox, the Arctic hare, and the ptarmigan.

The most identifiable Arctic animal is the polar bear, another year-round

Ptarmigans are brown in the summer. They turn white in the winter.

Think About It

Why do you think some Arctic animals turn white in the winter? How does this help protect them?

inhabitant. Polar bears have thick, whitish fur covering black skin. The fur and a thick layer of fat keep them warm. The black skin helps to absorb heat from the sun. Polar bears even have fur on the soles of their feet. Polar bears feed mainly on sea mammals, especially seals. Polar bears hunt both on the ice and in the water.

Polar bear cubs stick close to their mothers and learn how to survive in the Arctic tundra.

Alpine Animals

Animals in an alpine tundra must be able to survive high elevations, high winds, and difficult winters. There are lower levels of **oxygen** in an alpine tundra than there are in other biomes because there is less

Vocabulary

Oxygen is an element essential to all life on Earth. Oxygen is in the air animals breathe.

Marmots are rodents that hibernate during the tundra's cold winters.

oxygen in the air at high elevations. Many of the animals that live there have larger lungs to help them deal with the lower oxygen levels. Ibex, chamois, marmots, and many bird species make their home in an alpine tundra.

Mountain goats are a common animal in this biome, especially at higher elevations. Mountain goats are great climbers. They can climb on snow-covered cliffs and icy mountainsides. They also can jump more than 10 feet (3 m) in one leap.

This juvenile, or young, mountain goat has not yet grown its beard or horns!

Mountain goats' hair is coarse, shaggy, and white. Their coats protect them from cold mountain temperatures. Both males and females have a beard and horns.

People of the Tundra

Very few people live in the tundra. Arctic peoples, including the Eskimo (Inuit), tend to live in places where hunting and fishing can provide enough food year-round. The Eskimo are native people of the Arctic regions. They live in Greenland, Alaska, Canada, and eastern Russia. Europeans and others have called them Eskimo for hundreds of years. They have different names for themselves. In Canada and Greenland, they prefer to be called Inuit.

Inuit, Yupik, and Alutiit are some of the names Eskimo people call themselves.

In Alaska, most of them prefer the term Eskimo.

The Sami people live in a region called Lapland that stretches across Finland, Sweden, Norway, and parts of Russia. Much of this region is tundra. The Sami are known for herding reindeer. Traditionally, Sami are **nomadic**.

VOCABULARY

Nomadic people move from place to place. They do not stay in one town or village.

The landscape of some tundra regions has been greatly altered by humans. The construction of ski resorts, roads, and mines have affected the tundra. Hunting and oil drilling bring people to the tundra and also threaten the wildlife there.

This Sami man is dressed in traditional clothing while guiding a reindeer.

CHANGES ON THE TUNDRA

Human activity has made a huge impact on the animal and plant life in the tundra. Environmental scientists worry that more hunting, oil drilling, and other human activities will seriously damage the tundra.

Oil drilling has a big impact on the environment. This oil rig is in Krasnoyarsk Territory, Russia.

Global warming is the gradual increase

VOCABULARY

A **gradual** change happens slowly. The temperature of Earth did not go up all of a sudden. It has been slowly increasing for many years.

of the average surface temperature on Earth. This warming is caused by human activities, primarily the burning of fossil fuels. Global warming has already made noticeable changes in both Arctic and alpine tundra ecosystems.

The tundra is especially sensitive to global warming. The average global surface-air temperature has risen by approximately 1.5°F (0.9°C) since 1900. In the Arctic, average surface air temperatures have risen by 5.3° F (3.5°C) over the same period.

Human activity is the main contributor to global warming.

Global warming is making the ice sheets in the Arctic and Antarctica melt. This causes sea levels to rise all around the world. Coastal areas of the tundra have been affected by rising sea levels. The warmer temperatures have made it possible for tree species that previously could not survive in the tundra to start growing there. This crowds out the plants that grew there before. It is also a problem for the animals who depend on the native plants for food.

This photo of the Arctic tundra shows a lake that was formed by the melting of permafrost.

Additionally, with warmer

temperatures in the tundra, people are more likely to move into those areas and develop them. This will reduce the size of the habitat of animals native to those areas.

Think About It

Oil drilling and hunting are major threats to the tundra. What do you think the consequences of these actions are? What impact could they have on the tundra?

The shrinking of tundra habitats is a threat to snow leopards and other tundra species.

The Tundra's Future

The future of the tundra is at risk. The tundra plays a very important role particularly because of its permafrost. A significant amount of the world's carbon is frozen in the permafrost. If the permafrost thaws, that carbon will be released in the form of carbon dioxide and methane. These two gases are major causes of global warming.

The melting permafrost can cause strange shapes to show up on the surface of the tundra, as seen in this photo.

The effects of climate change are more noticeable in the tundra and are very worrisome to scientists. Many environmental scientists are studying the Arctic tundra and watching as ice continues to melt. This melting ice is contributing to rising sea levels.

Scientists have been trying to make more people aware of the problem of global warming. They hope that continued efforts to slow the warming may save the tundra and the plants and animals that live there.

THINK ABOUT IT

The tundra may seem far away, but climate change is a global problem. What can you do to reduce your impact on the climate?

Scientists hope to preserve the natural habitat of the plants and animals of the tundra.

GLOSSARY

algae Living things that make their own food but are less complex than plants.

blizzard A storm with strong winds and snow.

bog A wetland with wet, spongy soil.

carbon An important element that exists in all known living things.

carbon dioxide A colorless gas that is an essential part of Earth's atmosphere.

climate The weather in a certain place over a long period of time.

drainage The flowing away of water into the ground.

ecosystem All of the living and nonliving things in an area.

elevation How high a place is above sea level.

fossil fuels Fuels such as coal, oil, and natural gas that are the remains of things that lived long ago.

fungi A group of simple living things that are neither plants nor animals. Fungi include mushrooms, yeasts, and molds.

habitat The place or type of place where a plant or animal naturally or normally lives or grows.

hibernate To pass the winter in a sleeping or resting state.

lichen A fungus and an alga living together.

mammals Living things with a backbone that are warm-blooded, usually have hair or fur, and, if female, produce milk to feed their babies.

methane A colorless odorless gas that can be burned and is produced by breakdown of living things.

mosses Small, seedless plants that often grow in moist, shady places.

native Living or growing naturally in a particular region.

For More Information

Books

Cunningham, Kevin. *The Inuit* (True Books: American History). New York, NY: Scholastic, 2011.

Johansson, Philip. *The Tundra: Discover this Frozen Biome* (Discover the World's Biomes). Berkeley Heights, NJ: Enslow Elementary, 2015.

Marsico, Katie. *Polar Bear* (A Day in the Life: Polar Animals). Chicago, IL: Heinemann, 2012.

Sisk, Maeve T. *Arctic Foxes* (Animals That Live in the Tundra). New York, NY: Gareth Stevens Publishing, 2011.

Websites

DK Find Out!
https://www.dkfindout.com/us/animals-and-nature/habitats-and-ecosystems/tundra/

Kids Do Ecology: World Biomes
http://kids.nceas.ucsb.edu/biomes/tundra.html
http://kids.nceas.ucsb.edu/biomes/alpine.html

KidsKnowIt Network: Tundra Biome
https://kidsgeo.com/geography-for-kids/tundra-biome/
Facebook: @kidsknowit; Twitter: @KidsKnowIt

National Geographic: Tundra
https://www.nationalgeographic.com/environment/habitats/tundra-biome/
Facebook: @natneo; Twitter: @NatGeo

Index